# YOUR KNOWLEDGE HAS VALUE

- We will publish your bachelor's and master's thesis, essays and papers

- Your own eBook and book - sold worldwide in all relevant shops

- Earn money with each sale

Upload your text at www.GRIN.com
and publish for free

Thomas Gamsjäger

# Flow Cytometry of Intracellular Calcium in Platelets

GRIN Publishing

**Bibliographic information published by the German National Library:**

The German National Library lists this publication in the National Bibliography; detailed bibliographic data are available on the Internet at http://dnb.dnb.de .

This book is copyright material and must not be copied, reproduced, transferred, distributed, leased, licensed or publicly performed or used in any way except as specifically permitted in writing by the publishers, as allowed under the terms and conditions under which it was purchased or as strictly permitted by applicable copyright law. Any unauthorized distribution or use of this text may be a direct infringement of the author s and publisher s rights and those responsible may be liable in law accordingly.

**Imprint:**

Copyright © 2012 GRIN Verlag GmbH
Print and binding: Books on Demand GmbH, Norderstedt Germany
ISBN: 978-3-656-14662-9

**This book at GRIN:**

http://www.grin.com/en/e-book/190163/flow-cytometry-of-intracellular-calcium-in-platelets

**GRIN - Your knowledge has value**

Since its foundation in 1998, GRIN has specialized in publishing academic texts by students, college teachers and other academics as e-book and printed book. The website www.grin.com is an ideal platform for presenting term papers, final papers, scientific essays, dissertations and specialist books.

**Visit us on the internet:**

http://www.grin.com/

http://www.facebook.com/grincom

http://www.twitter.com/grin_com

Thomas Gamsjäger

# Flow Cytometry of Intracellular Calcium in Platelets

# Contents

Introduction ................................................................................................................. 3
The Role of Calcium in Intracellular Signal Transduction ........................................... 4
    Sources of Calcium ................................................................................................ 4
    Signal Transduction ................................................................................................ 4
    Activation of Platelets ............................................................................................. 6
        Adhesion ........................................................................................................... 6
        Shape Change .................................................................................................. 7
        Aggregation ...................................................................................................... 7
        Secretion .......................................................................................................... 7
        Procoagulant Activity ........................................................................................ 8
        Soluble Platelet Agonists .................................................................................. 8
    Inhibition of Platelets by cAMP ............................................................................... 9
Flow Cytometry ......................................................................................................... 10
    Principle ................................................................................................................ 10
    Parameters ........................................................................................................... 12
        Monoclonal Antibodies ................................................................................... 13
        Isotype Control ............................................................................................... 14
        Compensation ................................................................................................ 14
        Monoclonal Antibodies Used for the Flow Cytometric Characterization of Platelets ....... 15
        Intracellular Probes ........................................................................................ 16
    Presentation and Analysis of Data ....................................................................... 17
        Dotplots and Identification of Cells ................................................................. 17
        Further Analysis ............................................................................................. 18
        Regions and Gates ........................................................................................ 19
The Use of Fluo-3 as Intracellular Calcium Indicator ................................................ 20
    Brief Introduction .................................................................................................. 20
    Fluo-3 .................................................................................................................... 20
Application ................................................................................................................ 22
    Flow Cytometric Analysis ..................................................................................... 22
Conclusion ................................................................................................................ 25
References ............................................................................................................... 26
Abbreviations ............................................................................................................ 28
Index ......................................................................................................................... 29

# Introduction

During the last years flow cytometry has been developed into an important diagnostic tool with a wide range of applications. Whereas the differentiation of leukocytes is most often encountered, flow cytometry also renders itself especially useful in the study of platelets[1,2]. Commonly, surface markers are of key interest, but the fact that platelets can be conveniently activated in vitro offers the opportunity to investigate intracellular signal transduction processes, which lead to the release of calcium into the cytosol. And it is this rise of intracellular calcium that flow cytometry is able to detect in real time through the use of suitable calcium-sensitive markers.

It is the aim of the following chapters to put together an overview of this technique and therefore to provide a sound starting ground for its application to clinical and experimental investigations and also for its further development.

# The Role of Calcium in Intracellular Signal Transduction

Much evidence has been brought together supporting the concept that calcium is the key second messenger in intracellular signal transduction of platelets. The increase of cytosolic free calcium represents a pivotal step during activation[3]. It is accepted that the graded response of platelets to stimuli of varying intensity reflects a progressive increase in free calcium levels with a clear relationship between the intensity of the stimulus, the intracellular calcium concentration, and the intensity of the effects induced by the stimulus[4]. These effects, triggered at distinct calcium levels, include processes like shape change, adhesion, secretion and aggregation of the platelets.

## Sources of Calcium

The intracellular calcium concentration of resting platelets lies in the order of 40 to 100 nmol/l[3]. Compared to the calcium concentration in the extracellular environment (i.e. in blood plasma) of roughly 2.5 mmol/l this represents a very steep concentration gradient, which is maintained predominantly by the action of a $Ca^{2+}/Mg^{2+}$ ATPase pumping the calcium mainly into the dense tubular system and perhaps outside the cell[5].

Upon stimulation, the concentration increases to 2 to 10 µmol/l. Calcium is released from the main storage sites, namely the dense tubular system and the plasma membrane. The dense tubular system consists of membrane-limited tubules derived from the endoplasmatic reticulum of the megakaryocytes[4].

Depending on the agonist, considerable amounts of calcium can be brought to the cytosol from the extracellular fluid across the plasma membrane through specific calcium channels. Additionally, calcium is located in mitochondria and dense bodies (storage granules containing serotonin among other substances), but these sites are of little importance in the activation of the platelets[3].

Immediately after stimulation, the calcium concentration begins to revert to basal levels indicating that the calcium is redistributed to the storage sites, mainly the dense tubular system[3]. This is achieved by the aforementioned calcium pump, which is activated through phosphorylation by a cAMP-dependent protein kinase[6].

## Signal Transduction

Signal transduction pathways are shown in Figure 1. Platelet agonists like thrombin, ADP, collagen, and epinephrine, interact with their specific receptors on the cell surface. Agonist-receptor interaction stimulates the enzyme phospholipase C via G-proteins. The activated phospholipase C cleaves $PIP_2$ (phosphatidylinositol 4,5-bisphosphate) to $IP_3$ (inositol 1,4,5-trisphosphate) and DAG (diacylglycerol). While the water-soluble $IP_3$ enters the cytosol, the lipid-soluble DAG remains bound to the plasma membrane. Within milliseconds after receptor occupancy, $IP_3$ induces the release of calcium from the storage sites[7]. The mechanism of this release is not completely known, although some evidence exists that it is inhibited by calcium channel blockers[6].

In addition, the agonist-receptor interaction can lead to a rapid influx of calcium from the extracellular medium. It has been suggested that this process is governed by receptor-controlled calcium channels or alkalinization of the cytosol[7].

DAG, the second product of the cleavage of $PIP_2$ activates protein kinase C (PKC), which phosphorylates various proteins finally leading to the secretion of the contents of the storage granules. This DAG-PKC system can work independently from other signal transduction pathways, however, the efficiency is markedly enhanced by the presence of increased levels of free calcium[7].

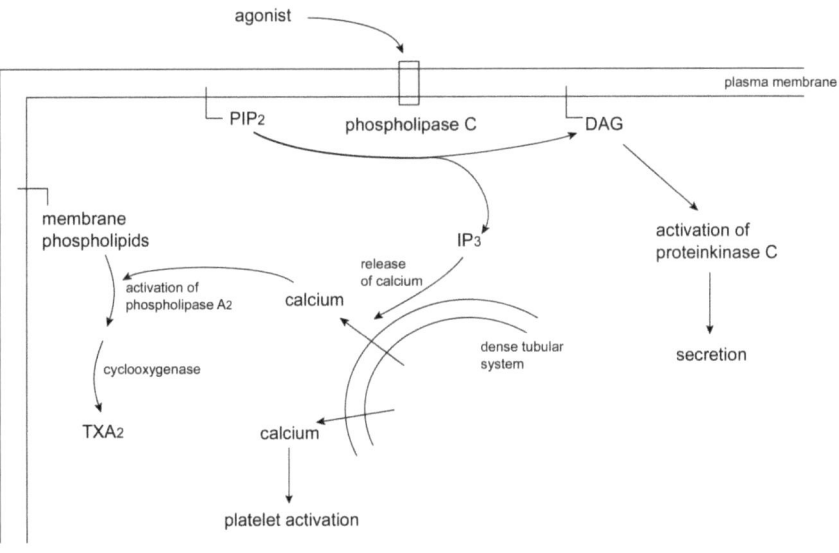

**Figure 1.** Principal pathways of signal transduction upon stimulation leading to the activation of platelets. $PIP_2$ - phosphatidylinositol 4,5-bisphosphate); $IP_3$ - inositol 1,4,5-trisphosphate; DAG - diacylglycerol; $TXA_2$ - thromboxane $A_2$.

High concentrations of cytosolic free calcium are not only able to prepare the platelets directly for the formation of a haemostatic plug but can also induce a further signal transduction pathway by activating phospholipase $A_2$. This enzymes uses membrane phospholipids like phosphatidylcholine and phosphatidylethanolamine as substrates for the formation of arachidonic acid, which is subsequently converted by cyclooxygenase to cyclic endoperoxides finally resulting in the generation of thromboxane $A_2$ (TXA$_2$). This substance is by itself a potent platelet agonist and induces vasoconstriction[8]. Inhibitors of the cyclooxygenase, like acetylsalicylic acid and other non-steroidal anti-inflammatory drugs (NSAID), reduce the formation of TXA$_2$. In the case of acetylsalicylic acid, the resulting decrease of platelet reactivity is used especially in the therapy of atherosclerotic diseases to reduce the risk of thrombus formation[9]. With regard to other NSAIDs this mechanism represents a potential side effect[10].


## Activation of Platelets


### Adhesion

Endothelial cells maintain a non-thrombogenic surface of the vessel walls to prevent the initiation of haemostasis. Damage to the endothelial cell layer due to mechanical injury or degenerative processes results in the exposure of the normally covered subendothelial matrix. The various constituents of the subendothelium (e.g. collagen) are highly thrombogenic in terms of both platelet and coagulation cascade activation[8].

Under conditions of high shear stress, which are present especially in arterioles (10 to 50 µm in diameter) and at stenoses of vessels, soluble von Willebrand Factor (vWF) adheres particularly to collagen and heparin-like glycosaminoglycanes. Initially, platelets interact with this form of bound vWF via the Gp Ib-IX complex, but as the high dissociation rate of this interaction is not able to bind the platelets firmly to the site of vessel injury, the platelets keep moving on the surface of the subendothelium in a rolling fashion, always maintaining the interaction between vWF and Gp Ib-IX. This results in a reduction of the velocity of the flowing platelets. The rolling process induces the activation of the platelets leading to a conformational change in the extracellular portions of glycoprotein IIb-IIIa (Gp IIb-IIIa) via the signal transduction pathways mentioned above[11].

Gp IIb-IIIa is a heterodimeric transmembrane molecule composed of an a and a b subunit ($a_{IIb}b_3$) belonging to the integrin receptor family together with several other glycoproteins[12]. The activated form of Gp IIb-IIIa is able to bind irreversibly to vWF completing the adhesion process of the platelet[11].

In parts of the circulation with low shear stress, the initiation of haemostasis relies at least on two different types of platelet-vessel wall interaction. Platelets bind to collagen mediated by the collagen receptor which is a heterodimer designated Gp Ia-IIa ($a_2b_1$). On the other hand, Gp IIb-IIIa receptors of *resting* platelets recognize and adhere to surface-bound fibrinogen which has undergone a conformational change, whereas soluble fibrinogen can only interact with Gp IIb-IIIa of activated platelets. Both mechanisms finally result in the activation of the platelets[11].

## Shape Change

In the course of platelet activation, the cells lose their normal (i.e. resting) discoid shape to appear as spheres with extending pseudopodia due to the increased polymerization of actin filaments and their association with myosin resulting in an increase of the surface area[4].

## Aggregation

Through mediation of soluble fibrinogen interacting with activated Gp IIb-IIIa receptors, platelets form larger aggregates, a process which is thought to occur mainly under conditions of turbulent flow providing intensive contact between the cells. Non-turbulent conditions are suggested to support the formation of a monolayer of cells[8].

Monoclonal antibodies like the antibody fragment c7E3 (abciximab) directed against Gp IIb-IIIa significantly reduce aggregation, which is now clinically used to prevent the formation of thrombi especially after coronary angioplasty procedures[13]. Studies have shown, however, that these antibodies are also able to reduce an agonist-induced calcium influx, suggesting that this glycoprotein may be involved in the calcium transport across the plasma membrane[6].

## Secretion

Apart from adhesion, shape change and aggregation another process, namely the secretion of storage granule contents, plays an important role in the formation of a haemostatic plug.

α-granules constitute one third to one half of the platelet granule population with a diameter of 300 to 500 nm. They contain numerous secretory proteins of which fibrinogen and vWF are the most important. Dense bodies, whose name is derived from their dense (i.e. dark) appearance on electron-microscopic imaging, provide especially ADP, calcium and serotonin[8].

Secretion occurs when the concentration of intracellular calcium reaches a certain level which is higher than the level necessary for the induction of shape change and the activation of Gp IIb-IIIa receptors[28]. The substances are released into the extracellular medium by fusion of the membrane-limited granules with the outer plasma membrane and partly with the open canalicular system, which represents invaginations of the plasma membrane. Once in the extracellular medium, the released substances contribute to the activation of further platelets and promote the overall process of haemostasis[4].

**Procoagulant Activity**

With regard to the coagulation system, platelets do not restrict themselves to the interaction with fibrinogen via their Gp IIb-IIIa receptors in order to form the haemostatic meshwork of cells and fibrin strands. Actually, platelets provide an ideal surface to enhance a number of steps in coagulation: By exposing negatively charged phospholipids during activation, platelets bind the constituents of the prothrombinase complex (factor Va, factor Xa and calcium) which catalyzes the conversion of prothrombin to thrombin; the tenase complex (factor VIIIa, factor IXa and calcium) promotes the activation of factor X; vWF, which interacts especially with Gp IIb-IIIa, is associated with factor VIII[8].

**Soluble Platelet Agonists**

In addition to the activation of platelets by interaction with the subendothelium, several soluble agents are able to stimulate platelets, most of them acting via G-protein-linked receptors.

ADP (adenosine diphosphate), which appears in plasma only through dense body secretion and lysis of blood cells, induces an increase in the cytosolic calcium concentration mainly via an influx of calcium across the plasma membrane and only to a lesser extent through the generation of $IP_3$ and the subsequent release of calcium from the dense tubular system. ADP is able to act synergistically with thromboxane $A_2$ ($TXA_2$). As described above (see section Signal Transduction), $TXA_2$ is produced intracellularly during platelet activation in the course of the arachidonic acid metabolism. To get into the extracellular fluid, it passes directly through the plasma membrane, where it supports the activation of further platelets by interacting with a specific $TXA_2$ receptor[11].

Thrombin, the generation of which is greatly amplified by platelets themselfes (see above), is a main product of the plasmatic coagulation cascade and converts fibrinogen to fibrin[14]. At the same time, thrombin is a potent platelet activator of central importance under physiological conditions. It induces the whole spectrum of activation effects and is able to act independently from the formation of TXA$_2$. Thrombin mobilizes calcium from internal and external stores[11].

Even though epinephrine activates platelets only at unphysiologically high concentrations, it is an interesting feature that the interaction of epinephrine with a$_2$ receptors of the platelet initiates activation only in the presence of extracellular calcium. The process seems to be completely dependent on an increased transmembrane permeability for calcium. Aggregation and secretion, but no shape change is induced[3,11].

## Inhibition of Platelets by cAMP

Whereas calcium is the pivotal intracellular messenger in terms of platelet activation, cAMP (cyclic adenosine monophosphate) plays a preeminent role in inhibiting platelet function[3].

Certain substances, especially the prostaglandins PGI$_2$ (also referred to as prostacyclin, mainly produced by endothelial cells) and PGE$_1$, inhibit platelet reactivity by raising the intracellular cAMP level through the stimulation of adenyl cyclase. Phosphodiesterase inhibitors exert a similar effect by reducing the activity of phosphodiesterase, an enzyme responsible for the degradation of cAMP.

cAMP decreases agonist binding to receptors, inhibits the generation of activating signal molecules in the phosphoinositide pathway and reduces the intracellular calcium concentration by activating a pump which sequestrates calcium from the cytosol mainly into the dense tubular system. These mechanisms result in the reduction of the magnitude and duration of calcium transients leading to decreased platelet reactivity. It is noteworthy that all physiologic agonists inhibit adenyl cyclase but the inhibition of this enzyme alone is not sufficient to trigger platelet activation[3].

# Flow Cytometry

Flow cytometry allows the (simultaneous) measurement of various cellular characteristics including size, granularity, expression of specific receptors and intracellular electrolyte concentrations.

Most previously available techniques to measure these parameters determined only the mean value for the whole population of cells. Flow cytometry, however, permits the measurement of these parameters in large numbers of single cells revealing possible heterogeneities within the cell population[15].

The following description applies especially to the analysis of platelets on a FACSCalibur™ flow cytometer (Becton Dickinson Biosciences, San Jose, CA, USA)[16,17].

## Principle

After preparation of the material of interest, the tube containing the sample is attached to the sample injection port of the cytometer. The pressure applied to the tube transports the sample at a rate of 12 to 60 µl/min into the flow chamber. Within this chamber, the sample fluid containing the cells is confined to the central portion of a flowing stream of sheath fluid (cell and particle free electrolyte solution), a principle that is known as hydrodynamic focussing. It is the aim to provide a condition of laminar flow which allows the cells to pass the observation point preferably in single file (Figure 2).

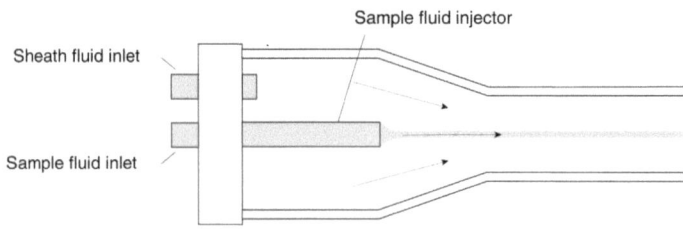

**Figure 2.** Schematic drawing of a flow chamber. Laminar flow conditions confine the sample fluid to the centre of the sheath fluid.

At the observation point, the passing cells are illuminated by an intensive beam of light. The source of light is typically an air-cooled argon ion laser emitting monochromatic (blue) light at a wavelength of 488 nm with a power of 15 mW.

The light emitted by the cells as a result of the illumination is collected by an optical system and finally registered in several detectors (Figure 3).

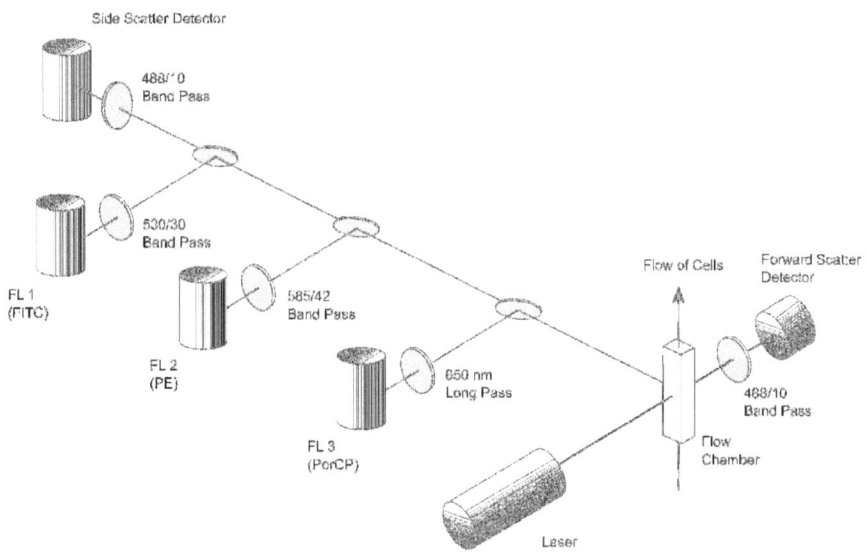

**Figure 3.** Principal layout of the optical system. Light from the flow chamber (scattered laser light and fluorescence) is lead to the detectors by using dichroic mirrors to split the beam. The detectors themselves are not able to register light of a specific wavelength. In order to analyse different colours separately, optical filters limit the access of the light to the detectors to defined ranges of wavelengths. Bandpass filters let light pass through between two values of wavelengths, blocking out lower and higher wavelength portions. They are described by their centre wavelength and their bandwidth. To give an example, light can penetrate a 530/30 bandpass filter only between 515 and 545 nm. On the other hand, a longpass filter, like the one used in a flow cytometer, allows light transmission above a defined value, in this case above 650 nm.

Forward scatter (FSC) detector: This is a photodiode which registers the relatively strong signal of laser light (488 nm) scattered by the cells due to their optical properties at a small angle from the incident laser beam.

Side scatter (SSC) detector (orthogonal scatter detector): A photomultiplier tube which detects scattered laser light at an angle of 90 degrees from the incident beam.

Fluorescence detectors 1, 2 and 3 (FL1, FL2, FL3): Upon illumination, cells stained with fluorescent probes emit light of a wavelength that is higher than the wavelength of the laser, a process called fluorescence (see section Parameters). The light collected at 90 degrees which consists of a spectrum of wavelengths (scattered laser light and fluorescence) is split into several ranges by the use of dichroic mirrors and filters. These distinct colours of the emitted light (green ~ 530 nm, orange ~ 575 nm, violet ~ 670 nm) are registered by three photomultiplier tubes (in addition to the side scatter detector mentioned above).

It is the task of the detectors to convert the incoming photons emitted by the flowing cells into a flow of electrons. Whereas photodiodes produce less than 8 electrons for every 10 photons, a photomultiplier tube is able to get a few hundred thousand electrons out of each photon, which makes these tubes especially well suited for the detection of the rather dim fluorescence light. The amplitude of the electrical current is proportional to the amount of light, one current pulse for each cell. Only when the height of the pulse reaches a preset threshold level, the pulse is passed on to the subsequent electronical processing.

After amplification in linear or logarithmic amplifiers, the signals are converted into digital form. Thereafter these digital data can be processed by a computer.

## Parameters

By means of the different detectors, the flow cytometer is able to gather information on several parameters, i.e. cellular characteristics.

Forward scatter (FSC): This parameter is a function of cell size. In a sample of whole blood, for instance, it allows to separate platelets from the larger red cells and leukocytes.

Side scatter (SSC; orthogonal scatter): This is a measure of the granularity of the cells as the light collected at 90 degrees is mainly scattered by intracellular membrane structures such as granules.

Fluorescence parameters: These parameters depend on the types of fluorescent probes used in a certain application. Two main types of such probes, monoclonal antibodies and intracellular dyes, will be discussed in more detail below.

Time: To monitor the dynamic changes of FSC, SSC and fluorescence over a certain space of time, it is possible to register time as an independent parameter during the acquisition of the signals emerging from the cells (kinetic measurements).

**Monoclonal Antibodies**

Monoclonal antibodies are immunological proteins (immunoglobulins) which are able to recognize one specific structure, i.e., one specific antigen. They are produced by forming hybrids between B-cells from animals immunized with an antigen and cultured cells derived from B-cell tumors[18]. For their use in flow cytometry, these monoclonal antibodies are covalently conjugated with fluorescent molecules[19].

Fluorescence is a process in which a molecule is able to absorb photons (for instance in the form of visible light) at a certain wavelength, which leads to an excited state, and to reemit photons at a similar wavelength allowing the molecule to get back to the electronic ground state[19].

The fluorescent substances (fluorochromes) are characterized by the absorption and emission spectra. The absorption spectrum defines the wavelengths at which the substance is able to take up the energy. It has to comprise the emission wavelength of the laser, usually 488 nm. The emission spectrum of the fluorescent label should preferably be located at some distance from the laser wavelength and determines which fluorescence detector has to be used for registration.

Important fluorochromes, which can be used in conjunction with 488 nm excitation, are fluorescein isothiocyanate (FITC), phycoerythrin (PE) and peridinin chlorophyll protein (PerCP)[19]- Figure 4.

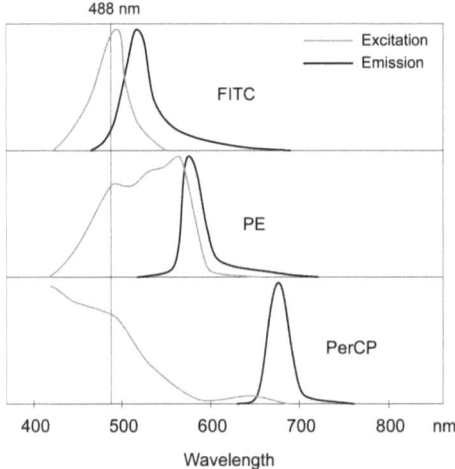

Figure 4. Absorption and emission spectra of fluorescein isothiocyanate (FITC), phycoerythrin (PE) and peridinin chlorophyll protein (PerCP). The absorption (excitation) characteristics of the fluorochromes (grey curves) are compatible with excitation by a 488 nm laser source. The emission spectra (black curves) are different between these dyes, allowing the simultaneous registration on the three appropriate photomultiplier tubes.

## Isotype Control

In immunofluorescence measurements it is necessary to define a threshold between cells which carry a specific antigen and other structures without this antigen. This is achieved by the use of a matched negative control. In the setup of the cytometer a sample is run which contains unspecific antibodies of the same isotype as the antibody used in the subsequent set of measurements, representing the background binding of antibodies independent of the target antigen. The amplifiers are adjusted in a way that the fluorescence of these unspecifically labeled cells is located below the arbitrary threshold of $10^1$ on a logarithmic scale (in the case of platelet studies, which exclusively employ logarithmic amplification). Per definition all cells in the subsequent measurements which are located above the level of $10^1$ are recognized as positive for the respective parameter[20].

## Compensation

Figure 5 shows that the emission spectra of some fluorescent dyes partly overlap. In multicolour analyses this would, if uncorrected, cause the registration of fluorescence of a certain colour by an inappropriate detector, i.e. a detector which is intended to measure only light of a different range of wavelengths, potentially leading to the misinterpretation of data from false positive or artefactual populations[21].

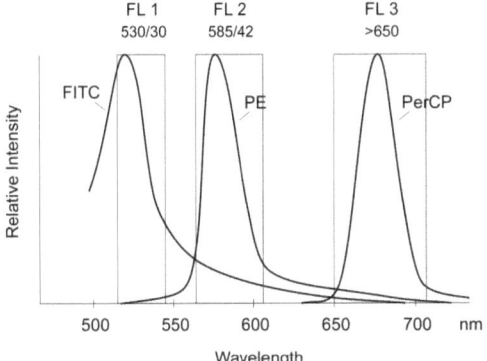

**Figure 5.** Spectral overlap. The figure depicts the emission spectra of FITC, PE and PerCP relative to the wavelength characteristics of their detectors (FL1, FL2 and FL3). The curves partly overlap, especially the curves of FITC and PE. Uncorrected, FITC would produce a considerably strong signal on FL2, which is intended to measure only PE emissions, whereas PE itself evokes a signal on FL1, the FITC detector. The same principle applies to all other combinations of fluorochromes necessitating electronic compensation to avoid distortion of the data.

Electronic compensation is the process of correcting for this spectral overlap by subtraction of unwanted signal. This is achieved by running cell samples in the setup of the instrument which are stained with only one fluorochrome at a time. The fluorescence of these cells will appear mainly on the appropriate detector, and with regard to the two remaining, inappropriate detectors the signals are defined as being identical to unstained cells.

**Monoclonal Antibodies Used for the Flow Cytometric Characterization of Platelets**

Monoclonal antibodies can be used to measure the expression of any platelet surface antigen[22]. However, the following antibodies are of particular interest:

Antibodies directed against CD61 allow to differentiate platelets from other blood cells. CD61 is a subunit of Gp IIb-IIIa (fibrinogen receptor) which is constitutively expressed on all platelets (resting and activated) and which does not appear on other (normal) blood cells[4].

The monoclonal antibodies against CD62P and the antibody named PAC-1 are able to detect activated platelets. This is the basis for the evaluation of the activity state of platelets in vivo and for the measurement platelet reactivity in vitro[22].

CD62P, also referred to as P-selectin, is an adhesion receptor which is normally located at the inner membrane of α-granules. Following stimulation, CD62P is exposed on the surface of platelets by fusion of the α-granules with the plasma membrane (a process called secretion or degranulation). There it plays an important role for the adhesion of leukocytes to platelets[22].

PAC-1 is an antibody which is able to recognize the activated form of the fibrinogen receptor (Gp IIb-IIIa), i.e. the fibrinogen receptor which has undergone a conformational change making it capable of binding fibrinogen[22,23].

Other important antibodies with regard to platelets are directed against Gp Ib (receptor for von Willebrand Factor; decrease upon activation)[22] and CD63 (a lysosomal membrane protein with increased expression on the platelet surface following activation-dependent degranulation of lysosomes)[24].

Applications include the detection of circulating activated platelets in acute coronary syndromes and in cerebrovascular ischemia, characterization of genetic defects of surface glycoproteins (like in Bernard-Soulier syndrome - deficiency in Gp Ib; Glanzmann´s thrombasthenia - deficiency in Gp IIb-IIIa), quality control of stored platelet concentrates, monitoring of platelet directed antithrombotic therapy[25], effects of extracorporeal circulation[26,27] or of drugs on platelets[28].

**Intracellular Probes**

Besides the characterisation of surface antigens, flow cytometry is also applicable to the measurement of intracellular parameters. Examples include the concentrations of electrolytes or messenger substances, pH and electrical charge.
The use of the intracellular probe Fluo-3 for the measurement of intracellular calcium will be described in the next chapter.

## Presentation and Analysis of Data[29]

**Dotplots and Identification of Cells**

An informative way to present data prior to more detailed analyses is achieved by using dotplots. These are two-dimensional diagrams in which each dot represents one cell. The position of an individual cell is defined by the values of the two parameters used as the axes of the diagram. Each combination of parameters acquired in the measurement can be used to create a dotplot diagram.

The combination of the parameters defining cell size (forward scatter, FSC) and granularity (side scatter, SSC) allows the initial evaluation and partly the identification of different cell populations.

The measurement of a whole blood sample leads to the appearance of two main clusters of cells. The population characterized by the lower values of FSC and SSC represents the platelets. Red blood cells and the small number of leukocytes form the second cluster and therefore it is not possible to distinguish between these two cell types in this form of diagram (Figure 6).

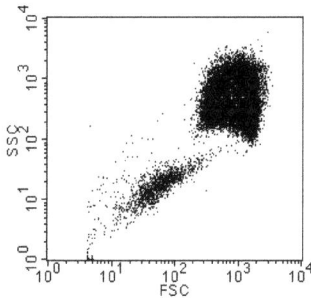

**Figure 6.** Dotplot FSC - SSC. The smaller population on the lower left represents platelets, the other population comprises leukocytes and mainly red blood cells. Note the logarithmic scale on both axes of the diagram.

Apart from using scatter characteristics, cells can by identified on the basis of fluorescence parameters. As mentioned above, platelets express constitutively CD61 on their surface. Thus, a fluorochrome-conjugated monoclonal antibody directed against this antigen allows the precise identification of the platelets (Figure 7). By the same principle it is possible to target other cell types, like leukocytes (using the pan-leukocyte marker CD45), or subpopulations of them.

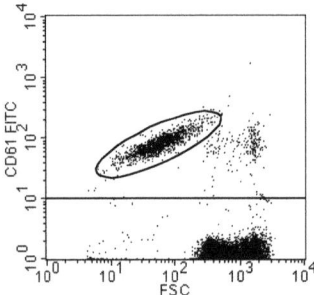

**Figure 7.** Dotplot FSC - CD61 FITC. Events above the arbitrary threshold of 101 (defined by using the appropriate isotype control) express enough CD61 to be positively identified as platelets.

**Further Analysis**

Another important way to present data employs histogram plots. A histogram is a one-dimensional frequency distribution of one of the acquired parameters. It offers a visual impression of the localization of the measured values on the respective scale and how these values deviate from the mean (Figure 8).

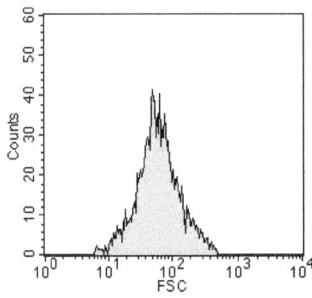

**Figure 8.** Histogram plot. It shows the frequency distribution of cells over the forward scatter.

In addition to the visual inspection of the plots, several statistical procedures can be performed on the acquired data. The most important of them are the calculations of the mean, geometric mean, median and standard deviation.

Two different approaches to data analysis may be used depending on the various target structures. One option is to determine the fraction of cells that is positive for a specific (activation-dependent) marker. This method allows to identify these activated cells with a high sensitivity on the basis of a predefined threshold between positive and negative fluorescence (determined using the isotype control). But it is not possible to quantify the exent to which antibodies are bound to the cells.

The latter can be achieved by the calculation of the mean fluorescence intensity which is the method of choice to determine minor changes in the surface antigen density[24].

**Regions and Gates**

Usually it is necessary to concentrate the analysis on a specific cell population leaving out all the other cells. For this reason the analytical software provides a tool which allows to set a region in a dotplot comprising the population of interest. A single region or the combination of several regions represent a gate which includes only the data of the cells under scrutiny and all subsequent analyses are performed selectively on the basis of these gated data (Figure 9).

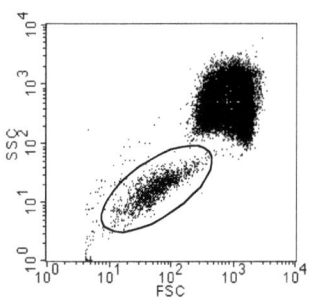

**Figure 9.** Dotplot FSC - SSC. The platelet population is encircled by an elliptical region allowing the specific analysis of these cells.

# The Use of Fluo-3 as Intracellular Calcium Indicator

## Brief Introduction

Until 1982, it was not possible to measure the intracellular calcium concentration in small intact cells. The measurements were restricted to large invertebrate cells, the dimensions of which allowed the use of microelectrodes. This limitiation was overcome by the introduction of Quin-2. This dye, which is loaded into cells in its acetoxymethyl (AM) ester form (see below), is able to bind cytosolic free calcium resulting in an increase in fluorescence. But its use is restricted by several disadvantages. The low extinction coefficient makes the detection difficult at low concentrations and at higher concentrations Quin-2 itself acts as a buffer for calcium[15].

Indo-1 represents a family of subsequently developed highly fluorescent calcium indicators. It shows large changes in the emission wavelength upon calcium binding. Use of the ratios of fluorescence intensities at two wavelengths allows the calculation of the calcium concentration independent of variability in cellular size or intracellular dye concentration. The most significant drawback of Indo-1 is its requirement for ultraviolet excitation[15].

Finally, Fluo-3 was introduced in 1989. It uses excitation with visible light and even though ratiometric determinations are not possible, this fluorescein-based dye is now extensively used in flow cytometry[15]. Other nonratiometric dyes with increased fluorescence on calcium binding include calcium green, calcium orange and calcium crimson, whereas Fura-Red exhibits decreased fluorescence on calcium binding[30].

## Fluo-3

In the absence of calcium, Fluo-3 is an essentially nonfluorescent substance which is loaded into cells in its acetoxymethyl (AM) ester form. The modification of carboxylic acids with acetoxymethyl ester groups produces an uncharged molecule which is able to permeate cell membranes. Once inside the cell, the lipophilic groups are cleaved by nonspecific esterases resulting in a compound which is trapped inside the cell due to the regained electrical charge[19].

As the large uncharged molecules of the AM ester, which is dissolved in anhydrous dimethylsulfoxide (DMSO), show a very low solubility in aqueous media, it is necessary to use the low-toxicity nonionic detergent Pluronic™ (Molecular Probes, Eugene, OR, USA) to assist the dispersion of the dye in the loading solution[19].

Fluo-3 exhibits an absorption spectrum that is compatible with excitation at 488 nm by argon ion laser sources (maximum absorption at 506 nm). The binding of calcium within the cytosol leads to a large increase in fluorescence intensity. Between resting cytosolic free calcium concentrations and indicator saturation, the enhancement is generally between 5- and 10-fold[19].

The emission characteristics, with the emission maximum at 526 nm, are virtually identical to those of FITC (fluorescein isothiocyanate), a fluorescent substance which is widely used in conjunction with monoclonal antibodies for immunophenotyping procedures[31]. Thus, Fluo-3 can be combined with many dyes used in multicolour analyses[30].

# Application

In the following section a working procedure is described, which has been applied in a clinical investigation[32].

Fluo-3 (Molecular Probes, Eugene, OR, USA; molecular weight 1130 Da) is prepared by dissolving 50 µg in 44.2 µl Pluronic™ (Molecular Probes; 10 per cent in dimethylsulfoxide DMSO) to yield a stock concentration of 1 mmol/l. The stock solution is stored at -20°C until the experiments are performed. Immediately before use, appropriate aliquots of Fluo-3 are diluted 1:10 in phosphate-buffered saline (PBS, Dulbecco´s, without calcium, magnesium and sodium bicarbonate, Life Technologies, Paisley, UK) to facilitate the handling of the otherwise very small volumes.

Aliquots of whole blood are diluted 1:10 with PBS and incubated with Fluo-3 at a final concentration of 5 µmol/l for 15 minutes at 37°C[33].

After the incubation, 25 µl of Fluo-3-loaded blood are diluted with 315 µl PBS in order to obtain a number of cells per µl suitable for flow cytometric measurements. In each experiment an additional sample of Fluo-3-loaded blood of the undiluted control is prepared in PBS for baseline calibration.

## Flow Cytometric Analysis

In the setup of the flow cytometer the amplification of the appropriate detector (FL1) is set to give a baseline fluorescence of approximately 2 arbitrary units.

A gate is set around the platelet population, which is identified by forward and side scatter characteristics (Figure 10).

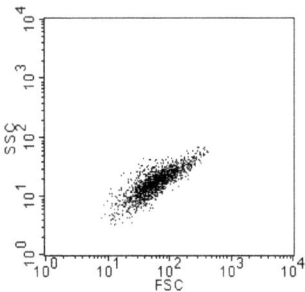

**Figure 10.** Dotplot FSC - SSC. The diagram shows only the platelet region, on which the analysis was based.

The fluorescence analysis is performed on the basis of this gate by using a dotplot with time on the x-axis and FL1-fluorescence on the y-axis. 15 consecutive regions in this dotplot allow the time-dependent interpretation of the data.

After determination of a baseline reading for approximately 10 seconds, the acquisition is briefly paused for the addition of the agonist (thrombin receptor activator peptide 6 (TRAP-6), final concentration 30 µmol/l; Bachem AG, Bubendorf, Switzerland) and immediately resumed afterwards (manipulation time approximately 10 seconds). Data are acquired for further 40 seconds.

Activation of the samples with TRAP-6 results in a marked increase in fluorescence intensity corresponding to an increase in the cytosolic free calcium concentration. Maximum values are reached within seconds after addition of the agonist and resumption of data acquisition, followed by a progressive decrease of fluorescence until the end of the measurement. Figure 11 depicts the typical development of the Fluo-3 intensity over time.

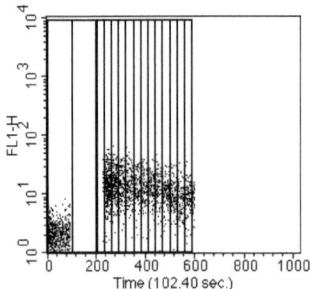

**Figure 11.** Dotplot Time - FL1. After baseline registration for about 10 seconds, the test tube was briefly removed from the sample injection port for addition of the agonist, and data acquisition was resumed immediately afterwards.

# Conclusion

Together with the humoral factors of the coagulation system platelets constitute an essential component in the process of haemostasis. Through steps including adhesion, shape change, aggregation, and secretion, platelets, which normally follow the stream of blood in a rather dormant state, are activated to form a haemostatic plug. These activation processes are triggered by external stimuli leading to a cascade of intracellular events with the release of calcium into the cytosol as their unifying feature.

Flow cytometry has been developed into an important investigational and diagnostic tool primarily for the detection of various surface markers of the different cells under scrutiny. But in addition, suitable dyes have been brought forward, which can be loaded into the cell, and exhibit a change in their fluorescent behavior upon contact with free calcium.

Against this backdrop flow cytometry offers a convenient way of investigating the degree of platelet activation through the measurement of the relative amount of cytosolic free calcium in vitro and in real time.

# References

1. Linden MD, Frelinger AL, Barnard MR, Przyklenk K, Furman MI, Michelson AD. Application of flow cytometry to platelet disorders. Semin Thromb Hemost 2004;30(5):501–11.

2. Michelson AD. Evaluation of platelet function by flow cytometry. Pathophysiol Haemost Thromb 2006;35(1-2):67–82.

3. Hawiger J. Signal transduction and intracellular regulatory processes in platelets. In: Colman RW, editor. Hemostasis and thrombosis: Basic principles and clinical practice. Philadelphia: Lippincott Williams & Wilkins; 1994.

4. Gawaz MP. Das Blutplättchen: Physiologie, Pathophysiologie, Membranrezeptoren, antithrombozytäre Wirkstoffe und antithrombozytäre Therapie bei koronarer Herzerkrankung. Stuttgart: Thieme; 1999.

5. White JG. Anatomy and structural organization of the platelet. In: Colman RW, editor. Hemostasis and thrombosis: Basic principles and clinical practice. Philadelphia: Lippincott Williams & Wilkins; 1994.

6. Holmsen H. Metabolism of platelets. In: Williams W, Beutler E, editors. Hematology. New York: McGraw-Hill; 1994.

7. Holmsen H. Platelet secretion and energy metabolism. In: Colman RW, editor. Hemostasis and thrombosis: Basic principles and clinical practice. Philadelphia: Lippincott Williams & Wilkins; 1994.

8. Kunicki TJ. Role of platelets in hemostasis. In: Rossi EC, Simon TL, Moss GS, Gould SA, editors. Principles of transfusion. Baltimore: Williams & Wilkins; 1996.

9. Hennekens CH. Update on aspirin in the treatment and prevention of cardiovascular disease. Am Heart J 1999;137(4 Pt 2):S9–S13.

10. Kenny GN. Potential renal, haematological and allergic adverse effects associated with nonsteroidal anti-inflammatory drugs. Drugs 1992;44 Suppl 5:31–6.

11. Harbrecht U. Die Thrombozytenaggregation: Physiologie und Biochemie. In: Hämostaseologie. Berlin: Springer; 1999.

12. Shattil SJ. Regulation of platelet anchorage and signaling by integrin alpha IIb beta 3. Thromb Haemost 1993;70(1):224–8.

13. Lefkovits J, Plow EF, Topol EJ. Platelet glycoprotein IIb/IIIa receptors in cardiovascular medicine. N Engl J Med 1995;332(23):1553–9.

14. Colman RW. Hemostasis and thrombosis: Basic principles and clinical practice. Philadelphia: Lippincott Williams & Wilkins; 1994.

15. Rabinovitch PS, June CH, Kavanagh TJ. Introduction to functional cell assays. Ann NY Acad Sci 1993;677:252–64.

16. BD FACSCalibur. 2011.

17. BD FACSCalibur Flow Cytometry System: Technical specifications. 2011.

18. Nelson PN, Reynolds GM, Waldron EE, Ward E, Giannopoulos K, Murray PG. Monoclonal antibodies. Mol Pathol 2000;53(3):111–7.

19. Haugland RP. Handbook of fluorescent brobes and research products. Molecular Probes; 2002.

20. O'Gorman MR, Thomas J. Isotype controls - time to let go? Cytometry 1999;38(2):78–80.

21. Roederer M. Spectral compensation for flow cytometry: visualization artifacts, limitations, and caveats. Cytometry 2001;45(3):194–205.

22. Michelson AD. Flow cytometry: a clinical test of platelet function. Blood 1996;87(12):4925–36.

23. Shattil SJ, Cunningham M, Hoxie JA. Detection of activated platelets in whole blood using activation-dependent monoclonal antibodies and flow cytometry. Blood 1987;70(1):307–15.

24. Schmitz G, Rothe G, Ruf A, et al. European Working Group on Clinical Cell Analysis: Consensus protocol for the flow cytometric characterisation of platelet function. Thromb Haemost 1998;79(5):885–96.

25. Michelson AD, Furman MI. Laboratory markers of platelet activation and their clinical significance. Curr Opin Hematol 1999;6(5):342–8.

26. Kestin AS, Valeri CR, Khuri SF, et al. The platelet function defect of cardiopulmonary bypass. Blood 1993;82(1):107–17.

27. Wahba A, Black G, Koksch M, et al. Cardiopulmonary bypass leads to a preferential loss of activated platelets. A flow cytometric assay of platelet surface antigens. Eur J Cardiothorac Surg 1996;10(9):768–73.

28. Kozek-Langenecker SA, Mohammad SF, Masaki T, Kamerath C, Cheung AK. The effects of heparin, protamine, and heparinase 1 on platelets in vitro using whole blood flow cytometry. Anesth Analg 2000;90(4):808–12.

29. Ormerod MG. Flow cytometry: A basic introduction. 2008.

30. June CH, Rabinovitch PS. Intracellular ionized calcium. Methods Cell Biol 1994;41:149–74.

31. Shapiro HM. Practical flow cytometry. New York: Wiley-Liss; 2003.

32. Gamsjäger T, Gustorff B, Kozek-Langenecker SA. The effects of hydroxyethyl starches on intracellular calcium in platelets. Anesth Analg 2002;95(4):866–9.

33. do Céu Monteiro M, Sansonetty F, Gonçalves MJ, O'Connor JE. Flow cytometric kinetic assay of calcium mobilization in whole blood platelets using Fluo-3 and CD41. Cytometry 1999;35(4):302–10.

# Abbreviations

| | |
|---|---|
| ADP | adenosine diphosphate |
| AM | acetoxymethyl |
| Ca | calcium |
| cAMP | cyclic adenosine monophosphate |
| CD | cluster of differentiation |
| Da | Dalton |
| DAG | diacylglycerol |
| DMSO | dimethylsulfoxide |
| FITC | fluorescein isothiocyanate |
| FSC | forward scatter |
| Gp | glycoprotein |
| $IP_3$ | inositol 1,4,5-trisphosphate |
| Mg | magnesium |
| NSAID | non-steroidal anti-inflammatory drugs |
| PBS | phosphate-buffered saline |
| PE | phycoerythrin |
| PerCP | peridinin chlorophyll protein |
| $PGE_1$ | prostaglandin $E_1$ |
| $PGI_2$ | prostaglandin $I_2$ |
| $PIP_2$ | phosphatidylinositol 4,5-bisphosphate |
| PKC | protein kinase C |
| SSC | side scatter |
| TRAP-6 | thrombin receptor activator peptide 6 |
| $TXA_2$ | thromboxane $A_2$ |
| vWF | von Willebrand Factor |

# Index

acetoxymethyl ester ..................................... 20
acetylsalicylic acid ........................................... 6
adenosine diphosphate ................................. 8
adenyl cyclase ............................................... 9
adhesion ........................................................ 6
ADP ..................................................... 4, 7, 8
aggregation ................................................... 7
alpha (α) granules ......................................... 7
arachidonic acid ............................................ 6
argon ion laser............................................. 11

Bernard-Soulier syndrome ........................... 16

$Ca^{2+}/Mg^{2+}$ ATPase.............................................. 4
calcium channel............................................. 4
calcium concentration, intracellular ................ 4
cAMP ............................................................. 9
CD61 ........................................................... 15
CD62P ......................................................... 16
CD63 ........................................................... 16
collagen ......................................................... 4
compensation ............................................. 14
cyclic adenosine monophosphate ................. 9
cyclooxygenase............................................. 6
cytosolic free calcium .................................... 4

DAG ............................................................... 5
dense bodies ................................................. 4
dense tubular system .................................... 4
detector, fluorescence ................................. 12
detector, forward scatter............................. 11
detector, side scatter ................................... 12
diacylglycerol ................................................. 5
dimethylsulfoxide.................................... 20, 28
DMSO ......................................................... 20
dotplot......................................................... 17

endoplasmatic reticulum ............................... 4

epinephrine................................................4, 9
fibrin.................................................................8
fibrinogen........................................................8
FITC ...............................................................13
flow cytometry ...............................................10
Fluo-3 ............................................................20
fluorescein isothiocyanate .......................13, 28
fluorescence...........................................12, 13
fluorochromes...............................................13
forward scatter..............................................12

gate ...............................................................19
Glanzmann´s thrombasthenia .......................16
glycoprotein IIb-IIIa..........................................6
G-protein ........................................................5

histogram plot...............................................18
hydrodynamic focusing.................................10

Indo-1 ............................................................20
inositol 1,4,5-trisphosphate ............................5
$IP_3$...................................................5, 8, 28
isotype control ..............................................14

matrix, subendothelial .....................................6
megakaryocytes .............................................4
mitochondria...................................................4
monoclonal antibodies..................................13

non-steroidal anti-inflammatory drugs .............6

PAC-1............................................................16
PE.................................................................13
PerCP............................................................13
peridinin chlorophyll protein....................13, 28
$PGE_1$ ................................................................9
$PGI_2$.................................................................9

phosphatidylcholine ......................................... 6
phosphatidylethanolamine .............................. 6
phosphatidylinositol 4,5-bisphosphate ........... 5
phosphodiesterase ......................................... 9
phospholipase $A_2$ ........................................... 6
phospholipase C ............................................. 5
phycoerythrin ........................................... 13, 28
$PIP_2$ ................................................................. 5
plasma membrane .......................................... 4
Pluronic$^{TM}$ ..................................................... 20
procoagulant activity ....................................... 8
prostacyclin ..................................................... 9
prostaglandins ................................................. 9

Quin-2 ............................................................ 20

region ............................................................ 19

sample rate ................................................... 10

second messenger ..........................................4
secretion ..........................................................7
serotonin ......................................................4, 7
shape change .................................................7
sheath fluid ....................................................10
side scatter ....................................................12
signal transduction .........................................4
stimulation ......................................................4
subendothelium ..............................................6

thrombin .....................................................4, 8
thrombin receptor activator peptide 6 ......23, 28
thromboxane $A_2$ ..........................................6, 8
TRAP-6 ..........................................................23
$TXA_2$ ...........................................................6, 8

von Willebrand Factor .............................6, 28

wavelength ....................................................12

30